I0088642

My Book

This book belongs to

Name: _____

Cover Design by :
Gowri Vemuri

First Edition :
May, 2019

Second Edition :

April, 2021

Author :
Gowri Vemuri

Edited by :
Raksha Pothapragada
Ritvik Pothapragada
Questions: mathknots.help@gmail.com

NOTE : VDOE is neither affiliated nor sponsors or endorses this product.

This book is dedicated to:

My Mom, who is my best critic, guide and supporter.

To what I am today, and what I am going to become tomorrow,

is all because of your blessings, unconditional affection and support.

This book is dedicated to the

strongest women of my life ,

my dearest mom

and

to all those moms in this universe.

G.V.

Visit www.a4ace.com

Also available more time based practice tests on subscription

The Virginia Board of Education and Virginia Department of Education (VDOE) have developed the Virginia Assessment Program (VAP) to measure and evaluate students' academic progress in the Standards of Learning (SOLs). The SOLs indicate Virginia's expectations for what students should know and be able to do in the subject areas of reading, writing, mathematics, science, and history/social science.

Students in grades 3-12 to take the Standards of Learning (SOL) assessments each year. Some of the tests are required by all students each year, and others are required only at specific grade levels. Additionally, with the removal of some SOL tests in recent years, the VDOE assigned the responsibility of the creation and administration of alternate, performance - based assessments on local divisions. Student scores from these tests determine a school's and the division's state accreditation and measures progress toward meeting federal targets.

Virginia Standards of Learning (SOL) tests are generally given online unless a student has an identified and documented need to be assessed using paper,pencil format. The test question format is typically multiple choice, and each test contains some technology enhanced items.

GRADE 3-8 STANDARDS OF LEARNING(SOL) TESTS

GRADE 3	GRADE 4	GRADE 5	GRADE 6	GRADE 7	GRADE 8
	VIRGINIA STUDIES				WRITING
MATH	MATH	MATH	MATH	MATH	MATH
READING	READING	READING	READING	READING	READING
		SCIENCE			SCIENCE

NOTE : VDOE is neither affiliated nor sponsors or endorses this product.

END OF COURSE STANDARDS OF LEARNING (SOL) TESTS

GRADE 9	GRADE 10	GRADE 11
ALGEBRA I	GEOMETRY	ALGEBRA II
EARTH SCIENCE	BIOLOGY	CHEMISTRY
WORLD HISTORY I	WORLD HISTORY II	VIRGINIA & US HSTORY
		WORLD GEOGRAPHY
		ENGLISH : READING
		ENGLISH : WRITING

Any Student taking one of the courses listed here is expected to take the corresponding end-of-course SOL test. The grade levels depicted here represent grade level at which students typically participate in these courses.

NOTE : VDOE is neither affiliated nor sponsors or endorses this product.

SOL Test Scoring and Performance Reports:

Standards of Learning assessments in English reading, mathematics, science and history/social science are made up of 35-50 items or questions that measure content knowledge, scientific and mathematical processes, reasoning and critical thinking skills. English writing skills are measured with a two-part assessment that includes multiple-choice items and an essay.

Student performance is graded on a scale of 0-600 with 400 representing the minimum level of acceptable proficiency and 500 representing advanced proficiency. On English reading and mathematics tests, the Board of Education has defined three levels of student achievement: basic, proficient, and advanced, with basic describing progress towards proficiency.

Performance Achievement Levels:

- The achievement levels for grades 3-8 reading and mathematics tests are: *Pass/Advanced, Pass/Proficient, Fail/Basic,* and *Fail/Below Basic.*

- The achievement levels for science tests, history tests, and End-of-Course (EOC) tests* are: *Pass/Advanced, Pass/Proficient,* and *Fail/Does Not Meet.*

- The EOC Writing (2010 SOL) test, EOC Reading (2010 SOL) test, and EOC Algebra II (2009 SOL) test have an achievement level of *Advanced/College Path* in place of the *Pass/Advanced* achievement level.

NOTE : VDOE is neither affiliated nor sponsors or endorses this product.

INDEX

www.a4ace.com   www.math-knots.com

FORMULA SHEET

1. Area of a triangle

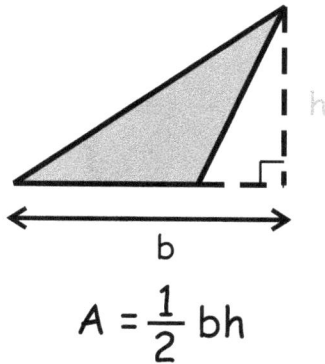

$$A = \frac{1}{2} bh$$

2. Area of a parellelogram

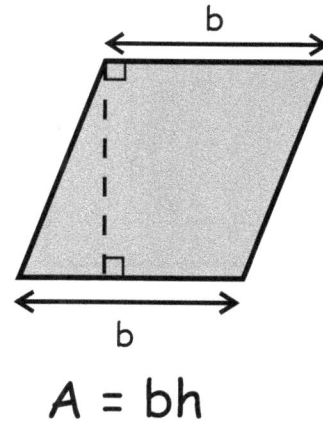

$$A = bh$$

3. Volume and Surface area of a cuboid

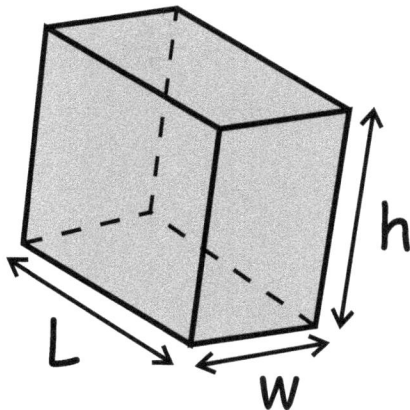

$$V = lwh$$
$$S.A = 2(lw + lh + wh)$$

4. Perimeter and Area of a Square

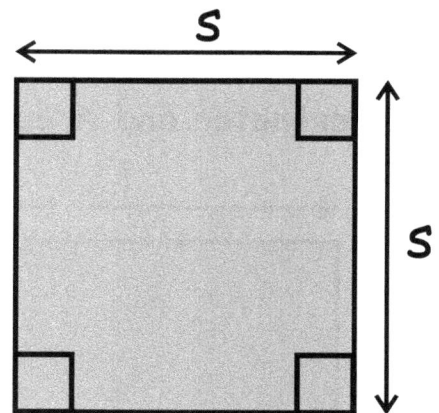

$$P = 4s$$
$$A = s^2$$

5. Area of a Trapezium

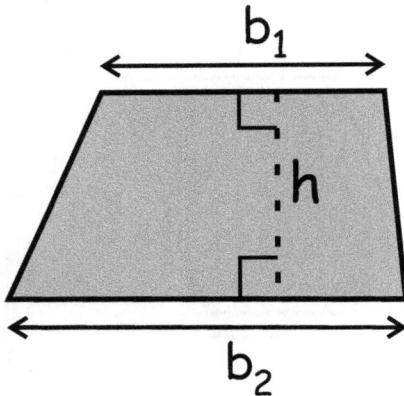

$$b_1$$

$$h$$

$$b_2$$

$$A = \frac{1}{2} h(b_1 + b_2)$$

6. Circumference and Area of a Circle

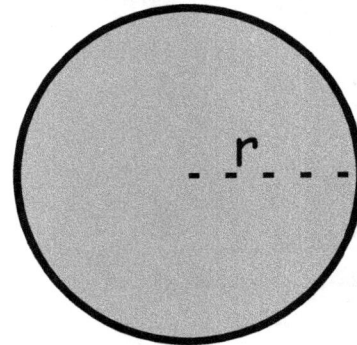

$$r$$

$$c = 2\Pi r$$
$$A = \Pi r^2$$

pi

$$\Pi = 3.14$$
$$\Pi = \frac{22}{7}$$

7. Perimeter and Area of a Rectangle

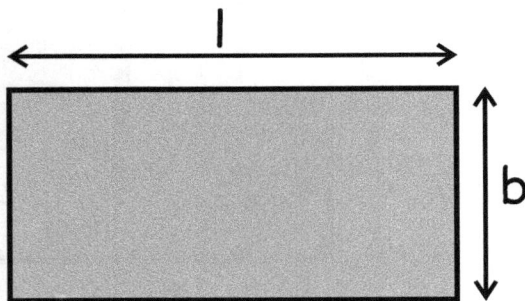

$$l$$

$$b$$

Area = l b ×
Perimeter = 2(l + b)

8. Acute angle

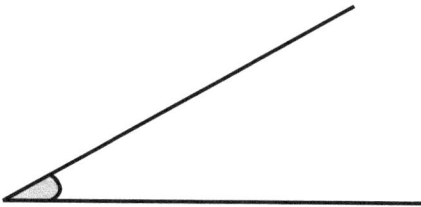

Any angle less than 90^0
is called as acute angled

9. Right angle

An angle equal to 90^0
is called as right angled

10. Obtuse angle

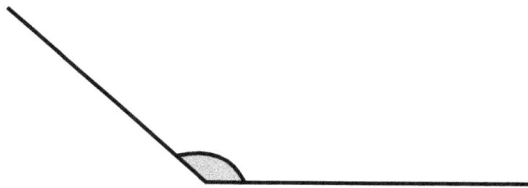

Any angle greater than 90^0
is called as obtuse angled

11. Perimeter of a polygon KLMNOP

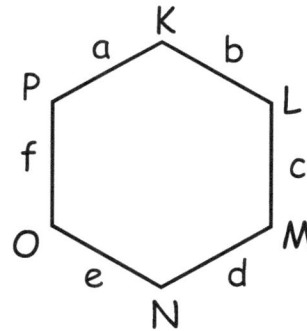

Perimeter =
$(a + b + c + d + e + f)$ units

Abbreviations

milligram	mg		
gram	g		
kilogram	kg		
milliliter	mL		
liter	L		
kiloliter	kL		
millimeter	mm		
centimeter	cm		
meter	m		
kilometer	km		
square centimeter	cm^2		
cubic centimeter	cm^3		

volume	V
total Square Area	S.A
area of base	B
ounce	oz
pound	lb
quart	qt
gallon	gal.
inches	in.
foot	ft
yard	yd
mile	mi.
square inch	sq in.

square foot	sq ft
cubic inch	cu in.
cubic foot	cu ft

year	yr
month	mon
hour	hr
minute	min
second	sec

Grade 5
SOL
Practice Test - 1

1. Evaluate $33.75 \div 5 = \boxed{?}$

    (A) 9.65   (B) 8.75   (C) 6.75   (D) 8.65

2. Joy spends $35 every week. How much did she spend after 41 weeks ?

    (A) 1435   (B) 1398   (C) 1450   (D) 1464

3. What is the sum of $3\dfrac{1}{8} + 2\dfrac{3}{2}$?

    (A) $1\dfrac{5}{8}$ (B) $3\dfrac{5}{7}$

    (C) $6\dfrac{5}{8}$ (D) $3\dfrac{3}{8}$

4. Human Habitats raised $7122 funds this year. Last year they have raised $6797. How much more did they raise this year than last year ?

    (A) 325   (B)329

    (C)328   (D)389

5. Dingy Dong town has two bridges of length 31.2 miles and 27.5 miles. What is the difference between the longest to shortest bridges in miles ?

    (A) 3.7   (B) 3.3.   (C) 3.5   (D) 3.6

6. Evaluate | 177777 ÷ 5 = ? |

    (A) 30555 R 4   (B) 3556 R 3

    (C) 35555 R 2   (D) 3505 R 5

7. Evaluate   957.027
                 + 101.951

    (A) 1058.978   (B) 1055.975

    (C) 1058.977   (D) 1057.977

8. Cathy orders two pizzas and slices each pizza into 8 pieces to
    distribute among 4 kids. How many pieces of pizza does each child
    got to eat ?

    (A) 8   (B) 6   (C) 2   (D) 4

9. Evaluate | $\frac{1}{5} - \frac{1}{9} = $? |

    (A) $\frac{4}{45}$ (B) $\frac{5}{8}$

    (C) $\frac{44}{45}$ (D) $\frac{46}{45}$

                  www.a4ace.com   www.math-knots.com

10. Evaluate $\boxed{9023 \div 3 = \boxed{?}}$

(A) 3001 R 0

(B) 3006 R 5

(C) 3007 R 3

(D) 3007 R 2

11. Estimate the sum of 0.7411 + 0.2589

(A) 0.55

(B) 1.35

(C) 0.912

(D) 1.0

12. Tracy organizes books in the library. She stacks 87 books in an hour. If she works for 4 hours in the library today. How many books did she stacked all together ?

(A) 344

(B) 548

(C) 348

(D) 7569

13. Which of the below fractions are ordered from greatest to least ?

(A) $\dfrac{21}{22}, \dfrac{22}{22}, \dfrac{15}{22}, \dfrac{19}{22}, \dfrac{7}{22}$

(B) $\dfrac{21}{22}, \dfrac{6}{22}, \dfrac{11}{22}, \dfrac{19}{22}, \dfrac{27}{22}$

(C) $\dfrac{21}{22}, \dfrac{16}{22}, \dfrac{15}{22}, \dfrac{11}{22}, \dfrac{7}{22}$

(D) $\dfrac{11}{22}, \dfrac{26}{22}, \dfrac{15}{22}, \dfrac{17}{22}, \dfrac{21}{22}$

www.a4ace.com   www.math-knots.com

14. What is the place value of 2 in 701.0255 ?

(A) Ones (B) Tenths

(C) Hundredths (D) Thousandths

15. Which of the below statements shows 0.921 written in words ?

(A) Nine hundred twenty-one thousandths
(B) Nine hundred twenty-one
(C) Nine and twenty-one thousandths
(D) Nine hundred twenty-one hundredths

16. $\dfrac{189}{1000}$ can also be written as ?

(A) 1.89 (B) 1.0089

(C) 18.9 (D) 0.189

17. Which number is three thousand sixty six and five thousandths ?

(A) 3066.555 (B) 36600.05

(C) 3660.05 (D) 3066.005

18. Which of the below statement is correct ?

    (A) 0.122 > 0.0122
    (B) 0.122 < 0.1022
    (C) 0.222 < 0.2120
    (D) 0.222 > 1.222

19. Find the value of 5 in 92.235 ?

    (A) Five tenths   (B) Five ones

    (C) Five thousandths   (D) Five hundredths

20. Round. 9087.763 to nearest hundredth ?

    (A) 9087.800   (B) 9087.760

    (C) 9088.763   (D) 9087.77

21. Which of the below angle measures $135°$?

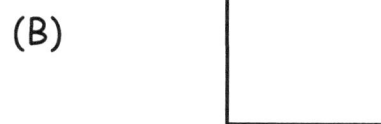

    (A)   (B)

    (C)   (D)

22. Anne has painted a nice picture She wants to frame it with a lace border. Given the dimensions of frame as 17 centimeters and 13 centimeters. Find the length of lace that she needs to buy ?

(A) 60 centimeters

(B) 221 centimeters

(C) 169 sq.centimeters

(D) 442 centimeters

23. What types of angles are used to form this triangle ?

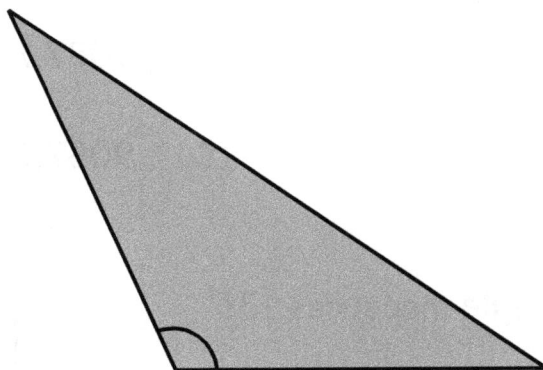

(A) Acute angle , Obtuse angle , Straight angle
(B) Acute angle , Right angle , Obtuse angle
(C) Acute angle , Acute angle , Acute angle
(D) Acute angle , Obtuse angle , Acute angle

24. Dora and Lora started cleaning their study room at 11:20 a.m. They finished cleaning by 3.03 pm. How much time it took for them to clean ?

(A) 4 hours, 43 minutes
(B) 3 hours, 43 minutes
(C) 3 hours, 40 minutes
(D) 4 hours, 33 minutes

25. Find the area of the square plot below with the side length of 31 miles ?

(A) 961 sq.miles   (B) 124 miles

(C) 62 Sq.Miles   (D) 50 Miles

26. Which of the below can be measured in cups ?

(A) Oranges   (B) Milk

(C) Bus   (D) Cereal

27. Which of the below picture shows the chord AB ?

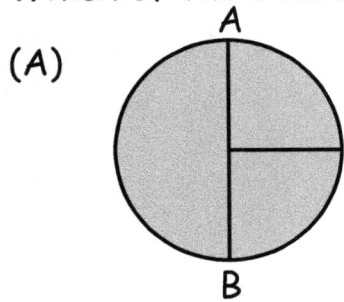

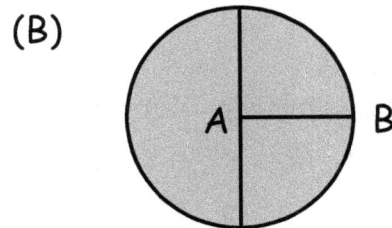

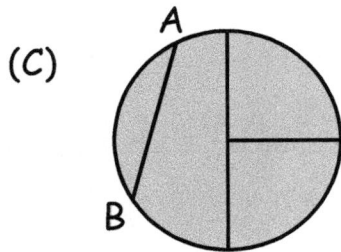

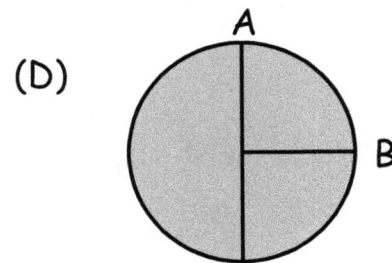

(A)

(B)

(C)

(D)

28. Which of the below picture has eight vertices , twelve edges and six faces ?

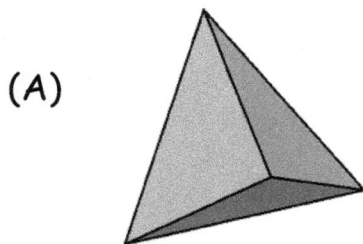

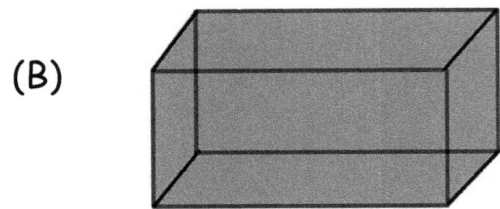

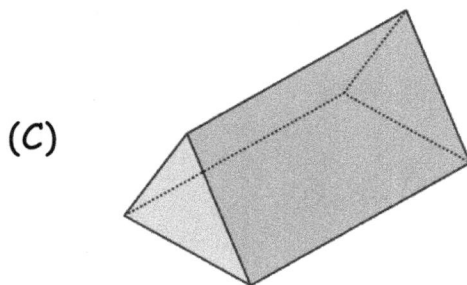

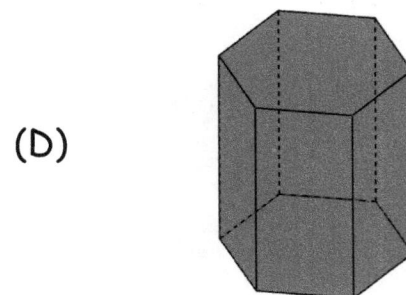

(A)

(B)

(C)

(D)

29. Which of the below pair of pictures are symmetric to each other ?

(A)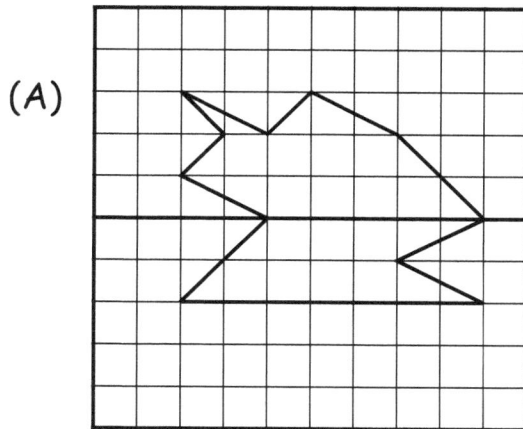

(B)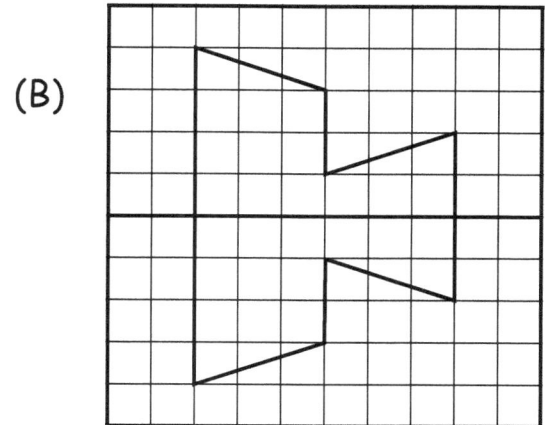

(C)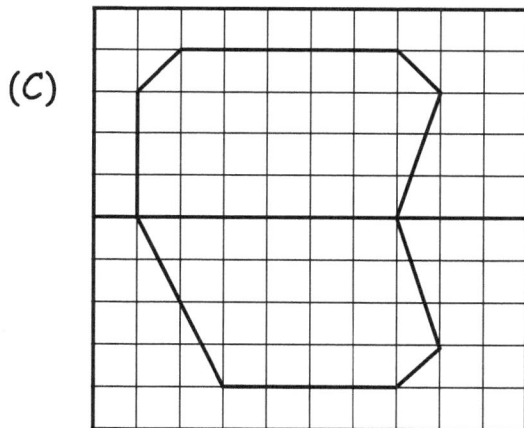

(D)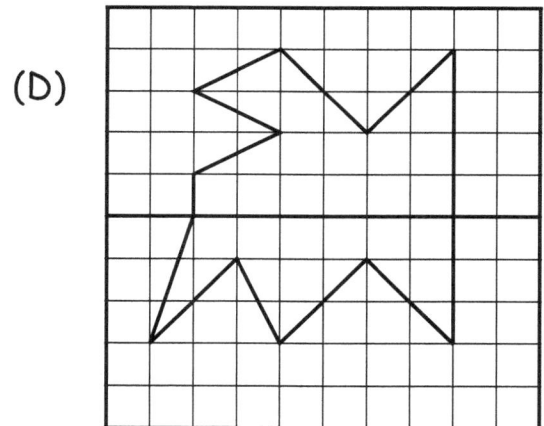

30. What is the probability of getting an even number when a 9 sided dice is rolled ?

(A) $\dfrac{7}{9}$

(B) $\dfrac{3}{8}$

(C) $\dfrac{5}{9}$

(D) $\dfrac{4}{9}$

 www.a4ace.com   www.math-knots.com

31. The Trivia quiz scores of fifth grade class are shown as stem-and-leaf plot as below. Based on the scores below, how many students have scored below 90 points ?

Stem	Leaf
5	0 , 0.8 , 9
6	5 , 5 , 5 , 6 , 9
7	3 , 6 , 6 , 8 , 9 , 9
8	3 , 3 , 8 , 9 , 9
9	0 , 0 , 1 , 1 , 3 , 3 , 5 , 5 , 5 , 8

Key	76
7	6

(A) 15   (B) 11   (C) 9   (D) 19

32. Below is sales data of various sandwiches at Deli stores this week. What is the mode of the data ?

121 , 155 , 131 , 131 , 101 , 150 , 144

(A) 131   (B) 155

(C) 101   (D)144

33. The John lake high school students voted for their one favorite book. The results are shown below

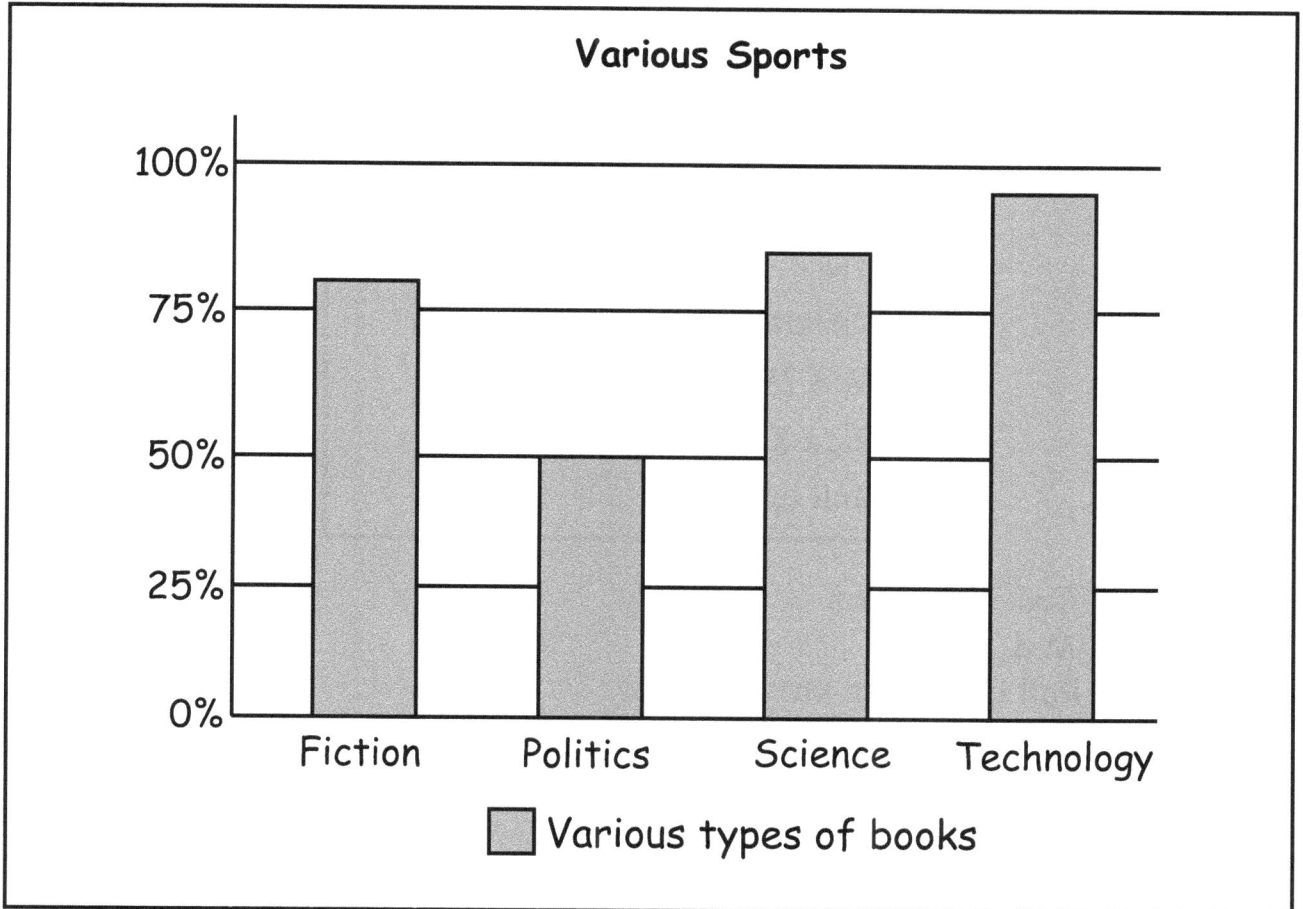

Various Sports

The greatest difference in the number of votes received was between the

(A) Fiction and Science

(B) Politics and Technology

(C) Fiction and Politics

(D) Science and Technology

www.a4ace.com   www.math-knots.com

34. Mrs.Tote is a pre-K teacher at Little Tots school. She places various treats in a surprise box. Each student picks 2 treats without looking at them. Based on the data in the table which of the below combinations is a possibility.

Counting Chips	
Color	Number in Box
Twix	1
Kitkat	9
M & M's	1
Rice Krisps	15
Roll up's	1

(A) 2 Twix,1 Kit-Kat
(B) 2 M & M's
(C) 1 Roll Up , 1 Kit-Kat
(D) 2 Roll Up's

35. Mary spins the arrow on the spinner shown.

What is the probability that the arrow will point to the section containing smileys on the first try ?

(A) 1 out of 8   (B) 1 out of 4

(C) 3 out of 7   (D) 3 out of 4

36. What is the variable in the below expression ?

$$3Y = -2019$$

(A) Y   (B) 2019   (C) Z   (D) X

37. Find the missing number in the pattern ?

$$11 , 17 , 23 , 29 , 35 , ? , 47 , 53$$

(A) 41   (B) 31   (C) 42   (D) 40

38. A number machine uses a rule to change numbers. The table shows the results.

In (x)	Out (Y)
7	14
8	16
11	22
27	54

(A) $\dfrac{x}{2} = y$ (B) $x + 7 = y$

(C) $2x = y$ (D) $\dfrac{y}{4} = x$

39. Aaron has 21 less cars than his friend Noah. If "C" represents the cars Noah had, which of the below equation represents, the number of cars Aaron has ?

(A) C + 13   (B) C - 21   (C) $\frac{C}{21}$ (D) 7*C

40. Emma has a magic machine. When Emma inputs the number in the Triangle, a different number comes out from square based on a rule. Find the math rule that magic machine uses.

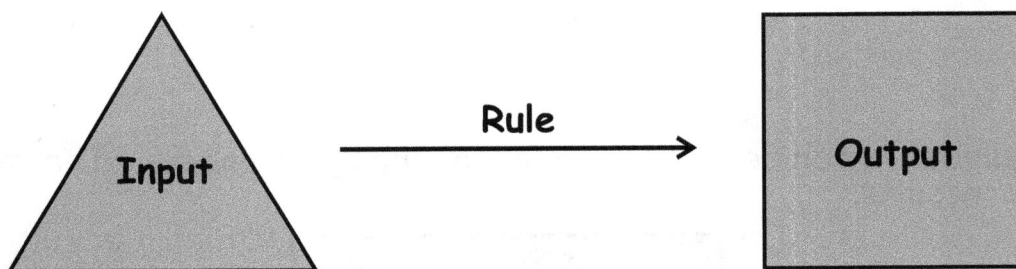

Input	Output
2	10
3	15
5	25

(A) Multiply by 5   (B) Multiply by 1

(C) Subtract 5   (D) Add 5

41. Find the missing number in the series

10000000 , 1000000 , ? , 10000 , 1000

(A) 100000   (B) 0100

(C) 1000   (D) 1100

42. Ben works at a store for $15 per hour. If he earns $180 this weekend. What does "h" represent in the equation 15h = 180

(A) The number of weeks Ben worked last month

(B) The number of days Ben worked last month

(C) The number of hours Ben worked last week

(D) The number of days Ben worked last week

43. Five less than twice a number, divided by three is written as ?

(A) $\dfrac{(2x - 5)}{3}$ (B) $(2x-5)*3$

(C) $\dfrac{7(2x + 5)}{3}$ (D) $\dfrac{(5 - 2x)}{3}$

44. Which of the following could be solved by using the open sentence
$$11 + b = 32$$

(A) Dan sells 11 bouquets every day for 32 days. How many bouquets did he sell all together ?

(B) Dan sells 11 bouquets yesterday. He sells more today .
All together he sells 32 bouquets in two days. What was the total. number of bouquets he sold today ?

(C) Dan sells 11 bouquets yesterday. He sells 45 more bouquets later today. At the end of the day, he had sold b bouquets. How many bouquet did he sell all together ?

(D) Dan sold 32 bouquets. If each bag had the same number of bouquets. How many bouquets were there in each bag ?

45. Which of the following can be solved using the open sentence
$$s - 15 = c \ ; \ s = 27$$

(A) Cathy has 15 fewer bags than Isabella. s represents the number of bags Isabella has, and c represents number of bags Cathy has. How many bags does Cathy have ?

(B) Cathy has 15 more bags than Isabella. If s represents the number of bags Isabella has, how many bags does Cathy have ?

(C) Isabella has 15 times more bags than Cathy. s represents the number of bags Isabella has, and c represents number of bags Cathy has. How many bags does Cathy have ?

(D) Cathy has 15 times more bags than Isabella. If s represents the number of bags Isabella has, how many bags does Cathy have ?

46. Students in Ms.Garry mathematics class tossed a six-sided number cube whose faces are numbered 1 to 6. The results are recorded in the table below.

Result	Frequency
1	3
2	6
3	4
4	6
5	4
6	7

Based on the above data , probability of getting a four is ?

(A) $\dfrac{6}{40}$

(B) $\dfrac{4}{30}$

(C) $\dfrac{8}{28}$

(D) $\dfrac{1}{5}$

47. Which set of data can be classified as quantitative ?
    (A) First names of students in a art club
    (B) Ages of students in a fifth grade class
    (C) Hair colors of students in a science club
    (D) Favorite sports of students in a soccer team

 www.a4ace.com   www.math-knots.com

48. Rita wants to pick a dress and its acceserries for her graduation party. She must choose 1 dress, 1 shoe pair and 1 Hand bag from the below.

Dress Combinations

Dress color	Shoe Color	Hand Bag
Blue dress	Black Shoes White Shoes Gold Shoes	Yellow Bag White bag

Based on the information in the chart, which tree diagram shows all of Rita's possible combinations ?

(A) Blue dress ⟶ Black Shoes ⟶ Yellow bag

(B)
Blue dress
- Black shoes → Yellow bag / White bag
- Gold Shoes → Yellow bag / White bag
- White shoes → Yellow bag / White bag

(C)
Blue dress
- Black shoes ⟶ Yellow bag
- White shoes ⟶ White bag

(D)
Blue dress ⟶ Gold shoes → Yellow bag / White bag

49. Which of the below options represents a translation of seven units to left and three units to down ?

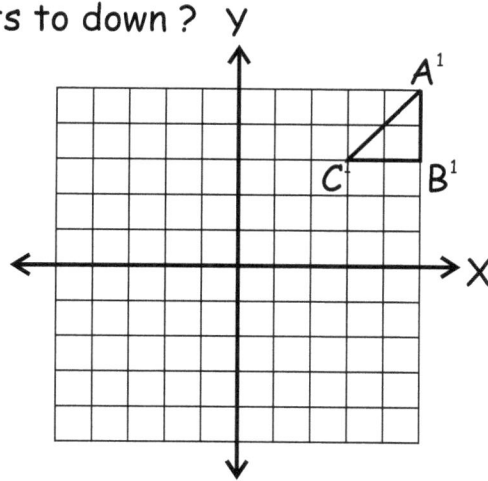

(A)

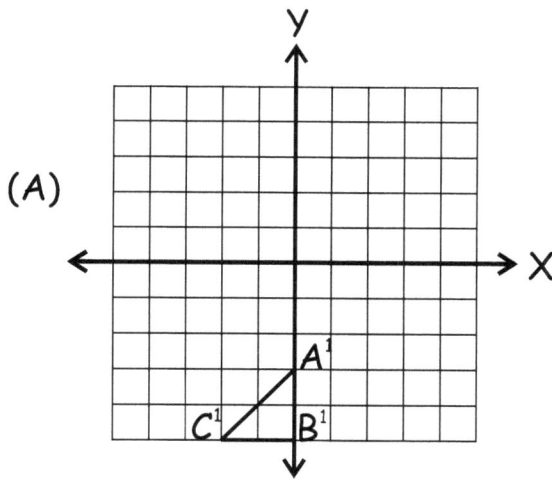

(B)

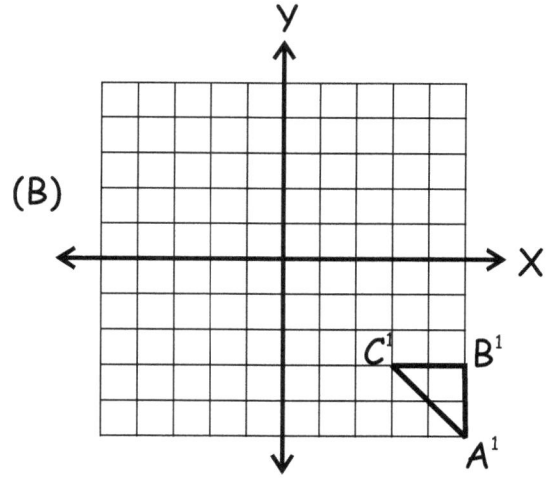

(C)

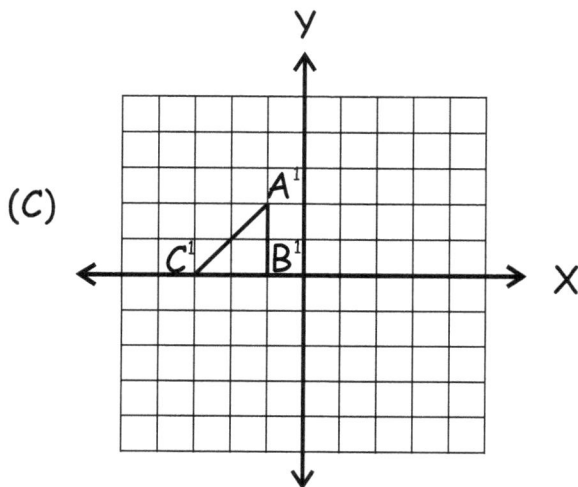

(D)

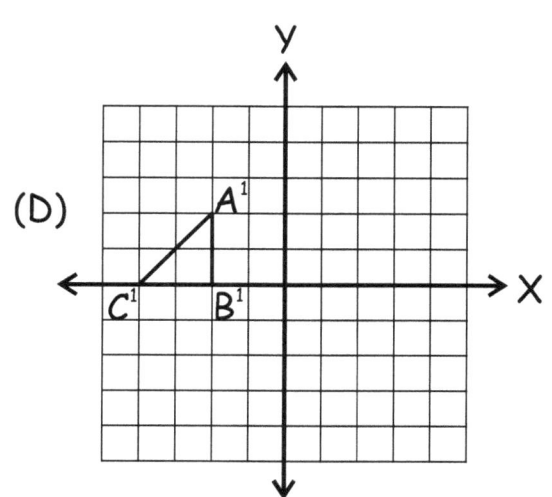

50. Find the reflection of the first figure across the dashed line ?

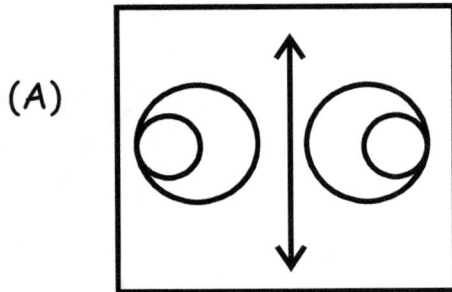

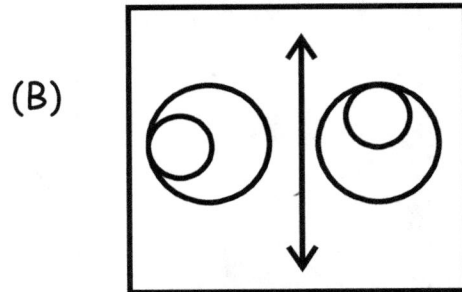

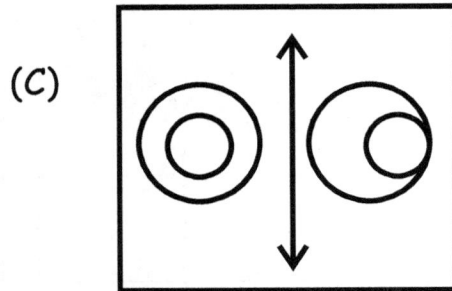

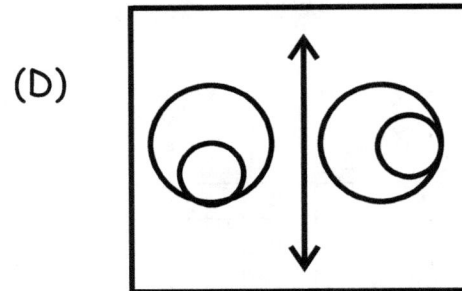

(A)

(B)

(C)

(D)

Grade 5
SOL
Practice Test - 2

1. Evaluate $\boxed{94.44 \div 6 = \boxed{?}}$

   (A) 15.75   (B) 15.74   (C) 15.76   (D) 15.77

2. Casey solves 11 math problems every day. How many problems did she have completed after 19 days ?

   (A) 190   (B) 191   (C) 189   (D) 219

3. What is the sum of $7\frac{1}{3} + 1\frac{3}{5} = ?$

   (A) $8\frac{14}{15}$ (B) $3\frac{14}{15}$

   (C) $6\frac{14}{15}$ (D) $3\frac{13}{15}$

4. Healthy Hearts marathon raised $997 funds this year. Their goal is to raise $2200. How much more money do they need to raise to reach their goal ?

   (A) 1204   (B) 1211

   (C) 1205   (D) 1203

5.   Cracky Crack town has two new streets of length 9.9 miles and 7.3 miles. What is the difference between the longest to shortest streets?

   (A) 2.7   (B) 2.6   (C) 17.2   (D) 17.6

6.   Evaluate $1111 \div 11 = \boxed{?}$

   (A) 101 R 7   (B) 101 R 0   (C) 101 R 2   (D) 101 R 1

7.   Evaluate   887.627
                 - 751.351

   (A) 136.2076   (B) 1036.277

   (C) 135.273   (D) 136.276

8.   Mary needs to pack 5 candies in each goodie bag for her daughter's birth day party. If her daughter invites 15 of her friends to the party. How many candies are needed in total for goody bags ?

   (A) 75   (B) 77

   (C)70   (D) 74

         www.a4ace.com   www.math-knots.com

9. Evaluate $\dfrac{1}{4} - \dfrac{1}{7} = \boxed{?}$

   (A) $\dfrac{4}{7}$ (B) $\dfrac{5}{28}$

   (C) $\dfrac{5}{28}$ (D) $\dfrac{3}{28}$

10. Evaluate $5949 \div 9 = \boxed{?}$

   (A) 661 R 0 (B) 662 R 5

   (C) 661 R 3 (D) 661 R 2

11. Estimate the value of 1.4511 + 2.3501

   (A) 3.5 (B) 3.9

   (C) 3.0 (D) 3.8

12. Jacks works at a gas station. He fills on an average 107 gallons of gas in an hour. If he works for 9 hours today, how gallons of gas did he fill in ?

(A) 960 (B) 997

(C) 963 (D) 11 R 8

13. Which of the below fractions are ordered from greatest to least ?

(A) $\frac{7}{9}, \frac{5}{9}, \frac{8}{9}, \frac{2}{9}, \frac{9}{9}$ (B) $\frac{7}{9}, \frac{4}{9}, \frac{3}{9}, \frac{8}{9}, \frac{1}{9}$

(C) $\frac{1}{9}, \frac{2}{9}, \frac{3}{9}, \frac{2}{9}, \frac{5}{9}$ (D) $\frac{7}{9}, \frac{5}{9}, \frac{3}{9}, \frac{2}{9}, \frac{1}{9}$

14. What is the place value of 3 in 111.0236 ?

(A) Ones (B) Tenths

(C) Hundredths (D) Thousandths

15. Which of the below statements shows 0.555 in words ?

(A) Five hundred Fifty-Five
(B) Five hundred Fifty-Five hundredths
(C) Five hundred Fifty-Five thousandths
(D) Five hundred Fifty-Five ten thousandths

16. $\dfrac{2221}{1000}$ can also be written as ?

  (A) 0.221 (B) 1.0221

  (C) 2.221 (D) 22.21

17. Which number is ninety nine and nine thousandths ?

  (A) 99.009 (B) 990.09

  (C) 990.009 (D) 99.0099

18. Which of the below statement is correct ?

  (A) 0.565 < 0.5065 (B) 0565 > 0.0565

  (C) 0.565 > 5.505 (D) 0.565 > 5.655

19. Find the value of 4 in 162.947 ?

  (A) 4 tenths (B) 4 ones

  (C) 4 tens. (D) 4 hundredths

20. Round. 0.1709 to nearest thousandth ?

(A) 0.2709   (B) 0.2000

(C) 0.1809   (D) 0.1710

21. Which of the below angle measures 55^0 ? (approximately)

(A) _____   (B)

(C)   (D)

22. Maya wants to plant tulips across the path of the triangular flower bed of equal sides. What is the length of the tulips flower bed ?

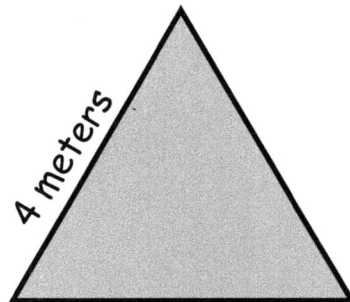

4 meters

(A) 12 centimeters   (B) 144 meters

(C) 72 meters   (D) 12 meters

23. What types of angles are used to form this triangle ?

    (A) Acute angle , Obtuse angle , Acute angle
    (B) Acute angle , Right angle , Acute angle
    (C) Acute angle , Acute angle , Acute angle
    (D) Acute angle , Acute angle , Right angle

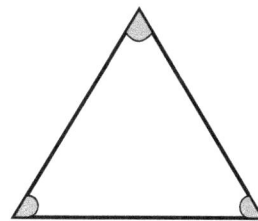

24. Dan takes a flight from Virginia at 6.06 am and reaches Mexico at 11:27a.m. What is his travel time ?

    (A) 6 hours, 21 minutes
    (B) 4 hours, 40 minutes
    (C) 5 hours, 37 minutes
    (D) 5 hours, 21 minutes

25. Find the area of the rectangular plot below with the side length of 11 and 12 miles ?

    (A) 132 sq.miles   (B) 133 miles

    (C) 46 sq.miles   (D) 264 miles

26. Find the reflection of the first figure across the dashed line ?

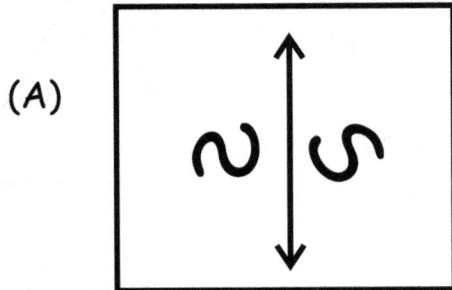

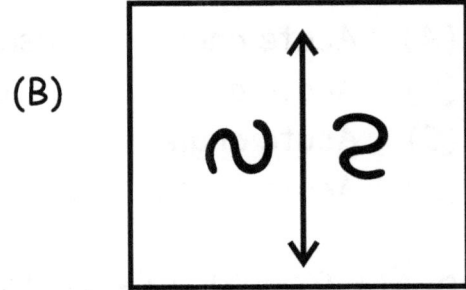

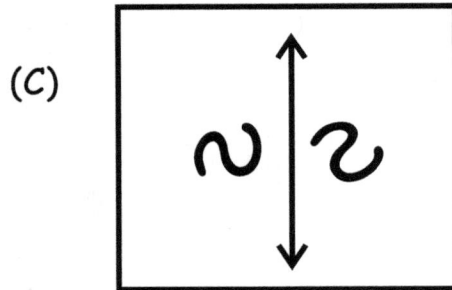

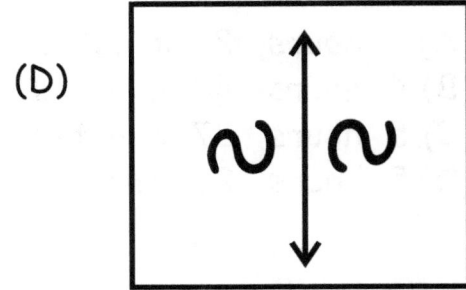

(A)

(B)

(C)

(D)

27. Which of the below picture shows the sector AOB ?

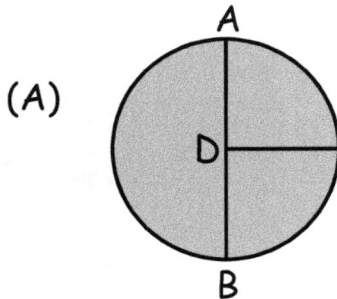

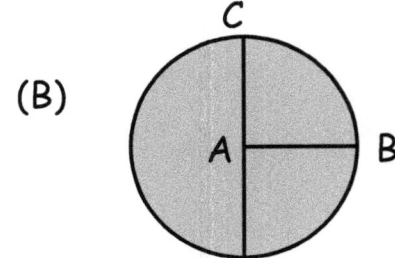

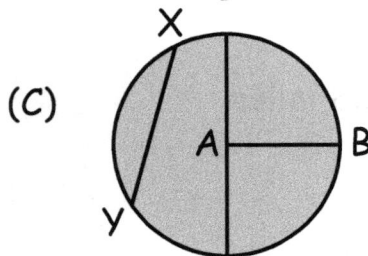

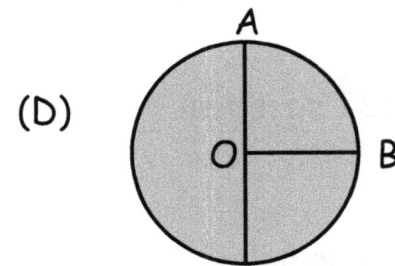

(A)

(B)

(C)

(D)

28. What is the probability of getting an odd number when a 6 sided dice is rolled ?

(A) $\dfrac{1}{6}$

(B) $\dfrac{3}{6}$

(C) $\dfrac{5}{9}$

(D) $\dfrac{2}{6}$

29. The Science scores of fifth grade class are shown as stem-and-leaf plot as below. Based on the scores below, how many students have scored 80 points and above ?

Stem	Leaf
5	0,1,8,9
6	2,3,3,6,9
7	3,5,0,7,8,9
8	2,3,7,4,2,7
9	0,1,2,2,3,4,4,7,6,8,0

Key	76
7	6

(A) 14

(B) 10

(C) 18

(D) 17

30. Below is the sales data of various ice-creams at school's cafe this week. What is the mode of the data ?

$$55 , 42 , 49 , 55 , 67 , 55 , 39$$

(A) 42   (B) 33

(C) 55   (D) 67

31. The Moses middle school students voted for their one favorite colors. The results are shown below.

Favorite color votes	
Red	69
Black	74
Blue	88
Pink	73
Orange	85
White	96

Which color received highest votes ?

(A) Red   (B)   White

(C) Blue   (D)   Orange

32. Mrs. Lara places various gifts in a surprise box for her son's birth day party. Each child in the party picks 3 gifts without looking at them. Based on the data in the table which of the below combinations is not a possibility.

Gift Item	Count
Red car	8
Green car	7
Black car	4
White car	3
Pinball game	10
Rubik's cube	2

(A) 3 Green cars

(B) 3 Rubik's Cube

(C) 3 Black Cars

(D) 1 Green Cars, 2 Pin the ball

33. Mary spins the arrow on the spinner shown.

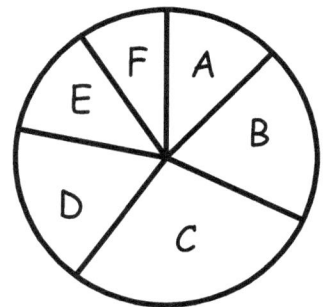

What is the probability that the arrow will point to the section containing "B" on the first try?

(A) 1 out of 8

(B) 1 out of 6

(C) 3 out of 6

(D) 5 out of 7

 www.a4ace.com   www.math-knots.com

34. What is the variable in the below expression ?

$$\frac{11011}{y} = 11$$

(A) y

(B) 11

(C) 1001

(D) X

35. Find the missing number in the pattern ?

10 , 25 , 40 , 55 , ? , 85 , 100

(A) 60

(B) 75

(C) 65

(D) 70

36. A number machine uses a rule to change numbers. The table shows the results.

(A) $(\frac{x}{2}) - 1 = y$

(B) $2x - 1 = y$

(C) $2x + 1 = y$

(D) $(\frac{y}{2}) - 1 = x$

In (x)	Out (y)
3	5
4	7
11	21
17	33

37. Evelyn makes twice the number of bracelets than her friend Mia. If b represents the bracelets Mia had, which of the below equation represents, the number of bracelets made by Evelyn.

(A) b + 2

(B) b - 2

(C) $\frac{b}{7}$

(D) 2b

38. Emma has a magic machine. When Emma inputs the number in the pentagon, a different number comes out from circle based on a rule. Find the math rule that magic machine uses.

InPut	OutPut
11	22
37	48
51	62

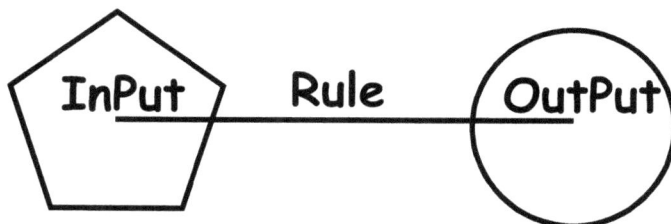

InPut   Rule   OutPut

(A) Multiply by 2

(B) Multiply by 11

(C) Subtract 11

(D) Add 11

39. Find the missing number in the series

$$1, 4, 9, ?, 25, 36$$

(A) 5   (B) 100   (C) 49   (D) 16

40. Sam works at a store for $23 per hour. If he earns $273 this weekend with a bonus of $43. What does "h" represent in the equation

$$23h + 43 = 273$$

(A) The number of weeks Sam worked last month

(B) The number of days Sam worked last month

(C) The number of hours Sam worked last week

(D) The number of days Sam is going to work next week

41. Eight more than a number squared, divided by seventeen is represented as ?

(A) $\dfrac{(x^2 + 8)}{17}$ (B) $(x^2 + 8) * 17$

(C) $\dfrac{(8 - x^2)}{17}$ (D) $(x^2 - 8) * 17$

42. Which of these could be solved by using the open sentence 9 + z = 32 ?

(A) Jack solves 9 puzzles in the morning for 32 days. How many puzzles did he solve all solve together ?

(B) Jack solves 9 puzzles yesterday. He solves 25 more puzzles later today. At the end of the day, he solves 32 puzzles. How many puzzles did he solve all together ?

(C) Jack solves 9 puzzles. If each day he had solved the same number of puzzles. How many puzzles did he solve ?

(D) Jack solves 9 puzzles yesterday. He solves more today . All together he solves 32 puzzles. How many puzzles did he solve today ?

43. Which of the following can be solved using the open sentence

g + 21 =m ; m = 70

(A) Mary has 21 more hair bands than Lucy. g represents the number of hair bands Lucy has, and m represents number of hair bands Mary has. How many hair bands does Lucy have ?

(B) Mary has 21 times fewer hair bands than Lucy. If g represents the number of hair bands Lucy has, how many hair bands does Mary have ?

(C) Mary has 20 times fewer hair bands than Lucy. g represents the number of hair bands Lucy has, and m represents number of hair bands Mary has. How many hair bands does Lucy have ?

(D) Mary has 12 fewer hair bands than Lucy. If g represents the number of hair bands Lucy has, how many hair bands does Mary have ?

44. Based on the table, what is the probability that a student chosen at random will vote for Larry ?

Name	No. of Supporters
Liam	28
Walsh	14
Tote	28
Ash	30

(A) $\dfrac{28}{100}$ (B) $\dfrac{30}{100}$

(C) $\dfrac{14}{100}$ (D) $\dfrac{0}{100}$

45. Which data set describes a situation that could be classified as quantitative ?

(A) The phone numbers in a directory

(B) The addresses for students at Kaiser Middle School

(C) The zip codes of residents in the city of Fairfax, Virginia

(D) The time it takes each of Miss. G's students to complete a test

46.   Which of the below can be measured in ounces ?

(A) Strawberries   (B) Milk

(C) Bus   (D) Liquid syrup

47.   Which of the below picture has six vertices , nine edges and five faces ?

(A) 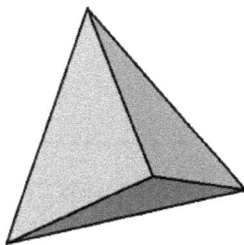   (B)

(C) 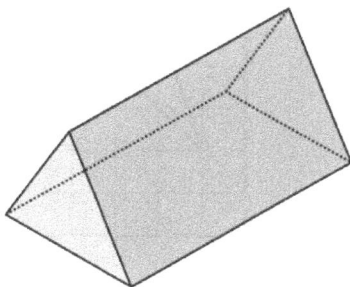   (D)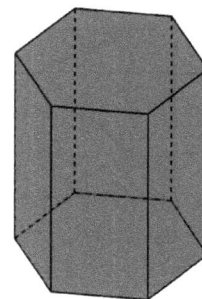

48. Which of the below pair of pictures are symmetric to each other ?

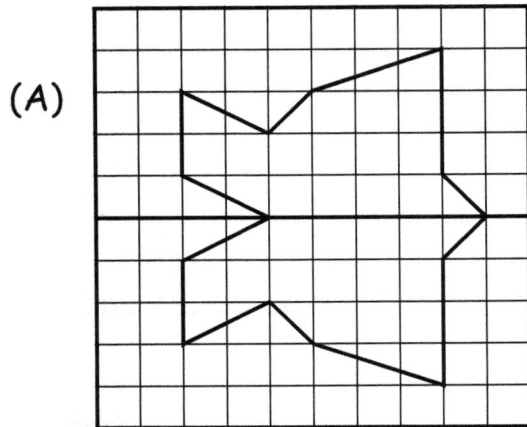

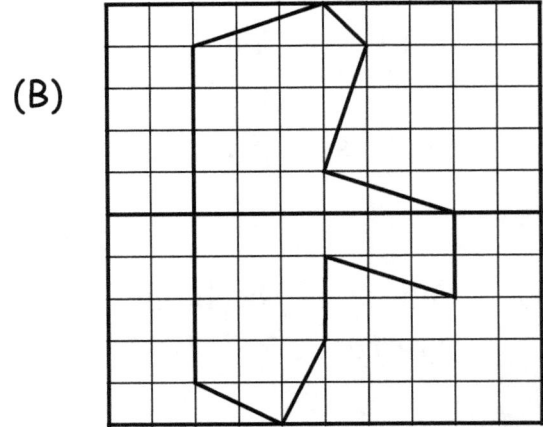

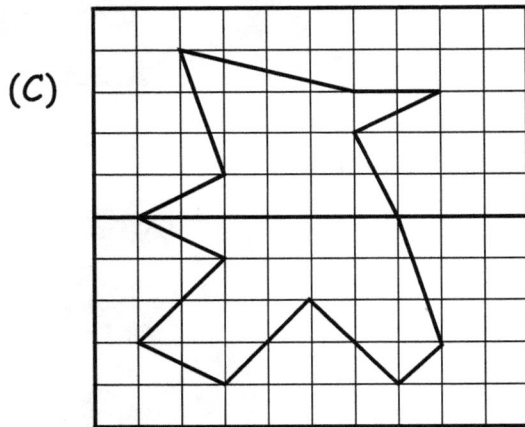

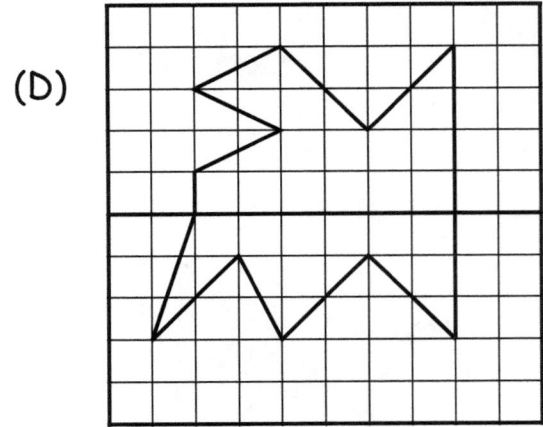

(A)

(B)

(C)

(D)

49. Which of the below options represents a reflection around x axis ?

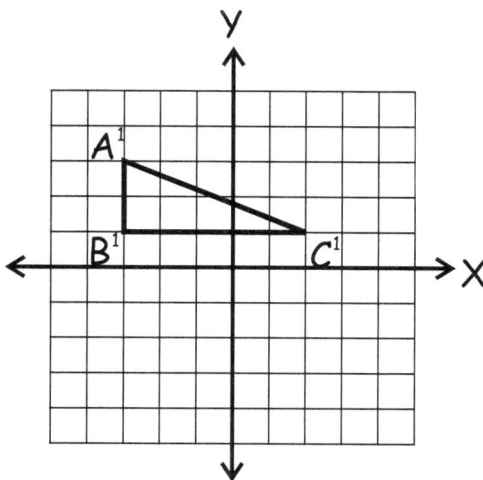

(A)

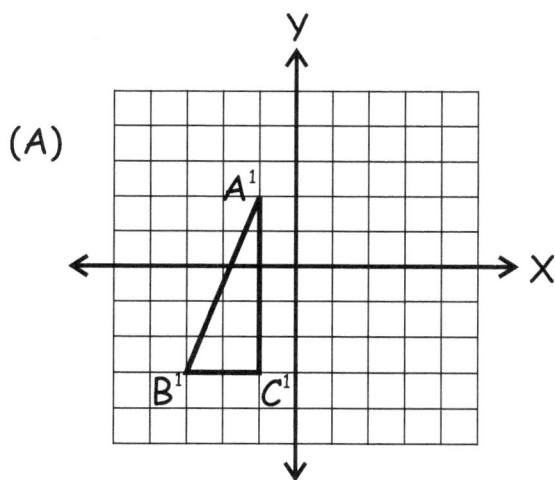

(B)

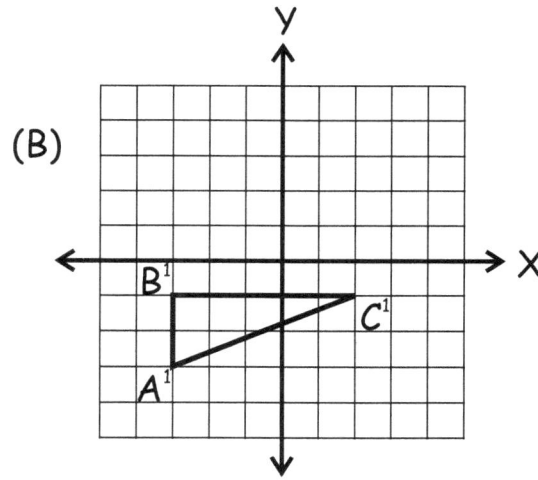

(C)

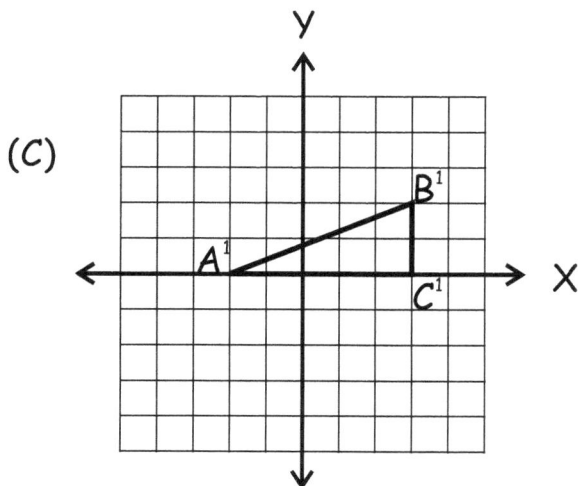

(D)

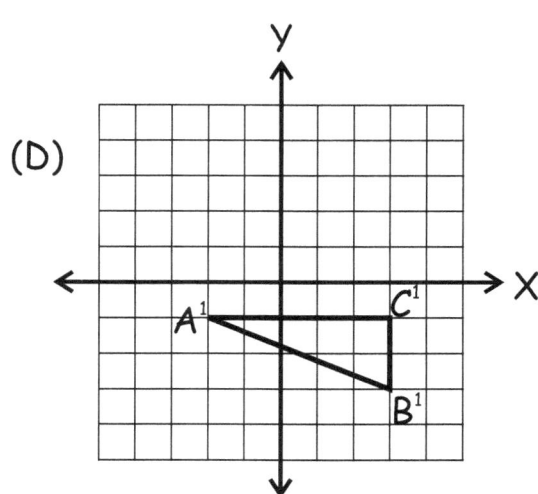

www.a4ace.com   www.math-knots.com

50. Stacy buys an ice cream She needs to chose one topping from each of the categories listed below.

Ice Cream Combinations

Ice Cream flavour	Toppings 1	Toppings 2
Vanilla	Sparkles Gummy bears	Chocolate syrup Caramel syrup

Based on the information in the chart, which tree diagram shows all of Stacy's possible combinations ?

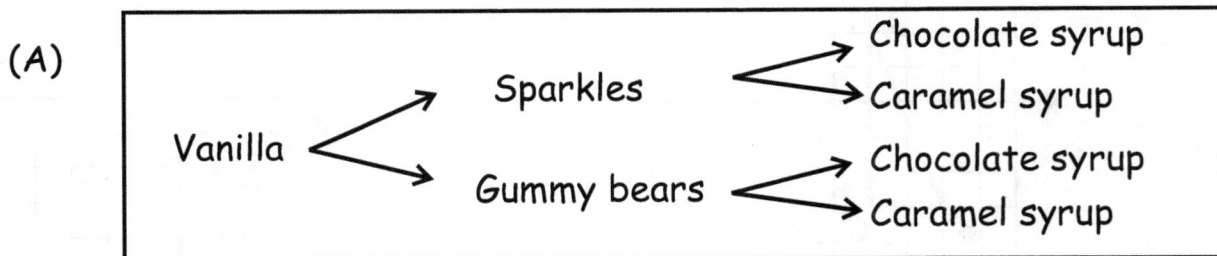

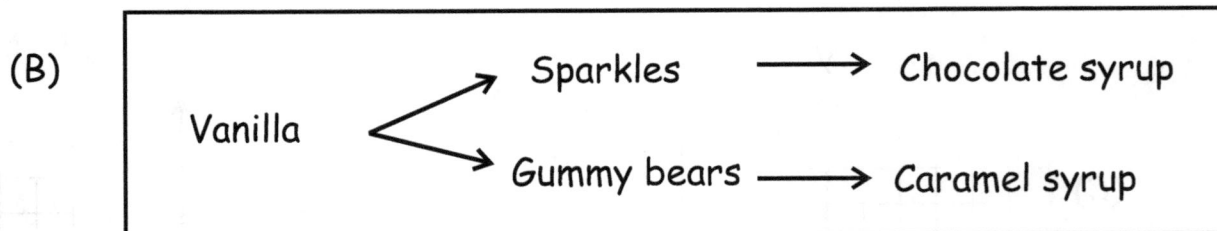

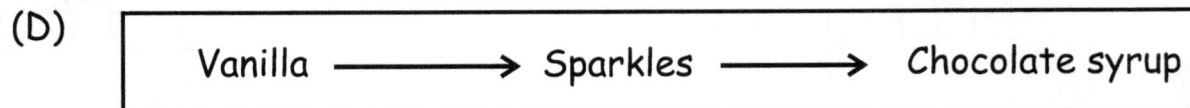

(A)

Vanilla
→ Sparkles → Chocolate syrup / Caramel syrup
→ Gummy bears → Chocolate syrup / Caramel syrup

(B)

Vanilla
→ Sparkles → Chocolate syrup
→ Gummy bears → Caramel syrup

(C)

Vanilla → Gummy bears → Chocolate syrup / Caramel syrup

(D)

Vanilla → Sparkles → Chocolate syrup

www.a4ace.com   www.math-knots.com

Grade 5
SOL
Practice Test - 3

www.a4ace.com

www.math-knots.com

1. Evaluate $5\sqrt{62500}$

   (A) 1050   (B) 1150   (C) 1250   (D) 1550

2. Evaluate $\dfrac{3}{5} + \dfrac{3}{2}$

   (A) $1\dfrac{1}{10}$ (B) $2\dfrac{1}{10}$ (C) $\dfrac{6}{10}$ (D) $3\dfrac{7}{7}$

3. What is the sum of $2\dfrac{3}{5} + 7\dfrac{1}{5} = ?$

   (A) $8\dfrac{3}{5}$ (B) $9\dfrac{4}{5}$

   (C) $6\dfrac{4}{5}$ (D) $3\dfrac{3}{5}$

4. Cathy bought 5 cases of Sparkles for July 4th. If each case has 21 boxes . How many Sparkle boxes did she purchased ?

   (A) 27   (B) 26

   (C) 115   (D) 105

                   www.a4ace.com   www.math-knots.com

5. Find the quotient of the below

$$28.84 \div 7 = \boxed{?}$$

(A) 4.12   (B) 41.02   (C) 4.0125   (D) 4.012

6. Evaluate $$110319 \div 11 = \boxed{?}$$

(A) 10029 R 0   (B) 10029 R 5   (C) 1029 R 0   (D) 1029 R 9

7. Evaluate

$$\begin{array}{r} 33.11 \\ \times \ 5.10 \\ \hline \\ \hline \end{array}$$

(A) 160.8061   (B) 166.851

(C) 168.861   (D) 165.551

8. Evaluate $$\dfrac{1155}{33} = \boxed{?}$$

(A) 32 R 32   (B) 35 R 35

(C) 35 R 0   (D) 34 R 29

www.a4ace.com   www.math-knots.com

9.   Evaluate

$$\frac{3}{7} + \frac{2}{7} = \boxed{?}$$

(A) $\frac{9}{7}$ (B) $\frac{1}{7}$

(C) $\frac{6}{7}$ (D) $\frac{5}{7}$

10.   Lola works two shifts at the book store this week end as per the below table

Games (Saturday Dec 22nd and Sunday Dec 23rd)	Starting and end time of each shift
Dec 22nd - Shift 1	9 AM to 12:30 AM
Dec 22nd - Shift 2	7 PM to 9:30 PM
Dec 23nd - Shift 1	8 AM to 12:30 PM
Dec 23nd - Shift 2	6 PM to 8:30 PM

How many hours did Lola worked this weekend ?

(A) 14   (B) 15

(C) 13   (D) 16

11. Evaluate

$$\begin{array}{r} 61.03 \\ - 27.79 \\ \hline \\ \hline \end{array}$$

(A) 33.25

(B) 33.24

(C) 32.33

(D) 33.33

12. Beth sold 829 cookies at school bake sale out of 911 cookies she made. How many more cookies does she need to sell ?

(A) 81

(B) 88

(C) 84

(D) 82

13. Which of the below decimal is equal to

$$\boxed{\dfrac{707075}{10000} = \boxed{?}}$$

(A) 70.7075

(B) 707.705

(C) 7.07075

(D) 0.707075

14. Which of the below fractions are ordered from least to greatest ?

(A) $\dfrac{5}{14}, \dfrac{6}{14}, \dfrac{3}{28}, \dfrac{5}{28}, \dfrac{1}{2}$

(B) $\dfrac{1}{2}, \dfrac{5}{28}, \dfrac{6}{14}, \dfrac{5}{14}, \dfrac{3}{28}$

(C) $\dfrac{3}{28}, \dfrac{5}{28}, \dfrac{5}{14}, \dfrac{6}{14}, \dfrac{1}{2}$

(D) $\dfrac{3}{28}, \dfrac{1}{2}, \dfrac{6}{14}, \dfrac{5}{14}, \dfrac{5}{28}$

15.   What is the value of 4 in 909.456 ?

(A)   Four tenths
(B)   Four
(C)   Four hundredths
(D)   Four tens

16.   Which of the below statement is correct?

(A) 12.032 < 12.302   (B) 17.002 > 17.502

(C) 67.99 < 87 .45   (D) 77.031 > 77.135

17.   Round 801.942 to the nearest hundredths

(A)   802.00   (B)   801.95

(C)   801.9   (D)   801.940

18.   Which of the below statements shows
        887.742 written or read in words as ?

(A)   Eight hundred eighty-seven and seven hundred fourty-two
        thousandths

(B)   Eight hundred and seven hundred fourty-two thousands

(C)   Eight hundred and eighty seven , fourty-two hundredths

(D)   Eight hundred and seven hundred fourty-two hundreds

19. Evaluate $\dfrac{987}{1000} = \boxed{?}$

    (A) 9.87

    (B) 0.00987

    (C) 0.0987

    (D) 0.987

20. 9987.786 round to near thousandths ?

    (A) 9987.8

    (B) 9987.79

    (C) 9987.7

    (D) 9987.78

21. Which of the below angle shows an acute angle ?

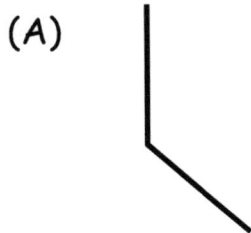

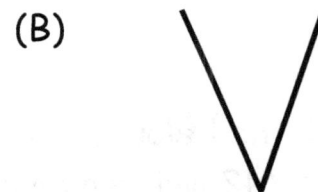

    (A)

    (B)

    (C)

    (D)

www.a4ace.com   www.math-knots.com

22. Which of the below figures has no sides ?

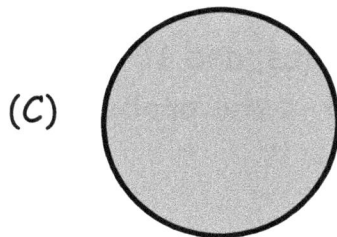

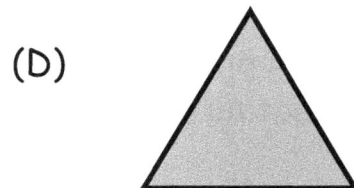

(A)

(B)

(C)

(D)

23. Find the reflection of the first figure across the dashed line ?

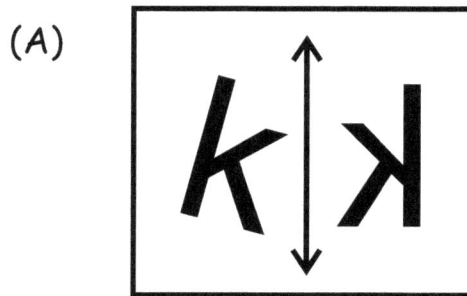

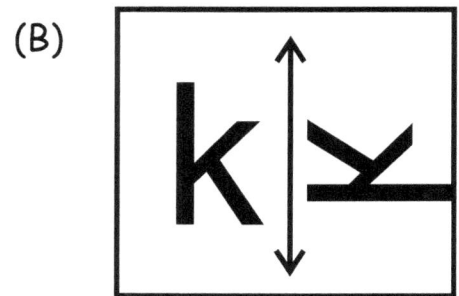

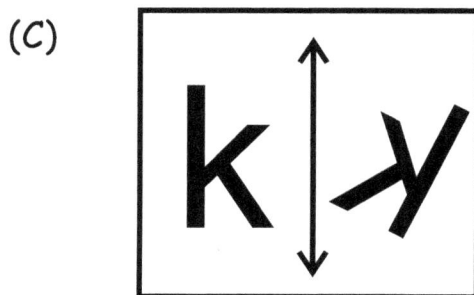

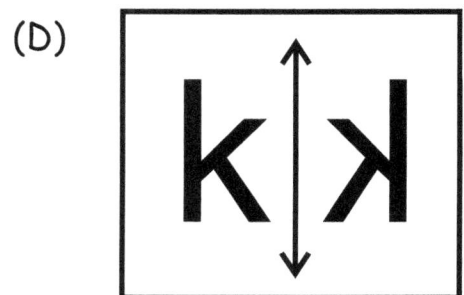

(A)

(B)

(C)

(D)

24. Rosy the architect draws a map of new shopping complex, garden around it and the parking spaces. After drawing she wants to measure the area of the building. She needs which of the below to measure the distance?

    (A) Compass   (B) Thread

    (C) Ruler   (D) Balance Scale

25. A box contains 9 green marbles, 5 White marbles, and 11 Yellow marbles. Diana picks up one marble from the box. What is the probability of selecting a Yellow marble ?

    (A) $\dfrac{11}{25}$ (B) $\dfrac{9}{25}$

    (C) $\dfrac{1}{25}$ (D) $\dfrac{1}{5}$

26. Terry works at a grocery stores from 10:15 A.M to 5.00 P.M.
    If he takes lunch break for half hour in between, How many hours did terry work at the grocery stores today ?

    (A) 6 hours 15 minutes

    (B) 6 hours 45 minutes

    (C) 5 hours 45 minutes

    (D) 6 hours 30 minutes

27. Which of the below pair of pictures are symmetric to each other ?

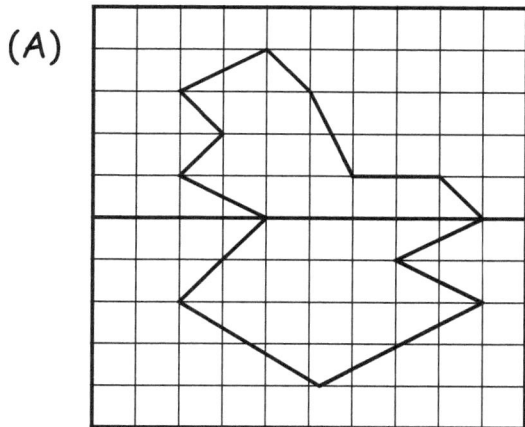

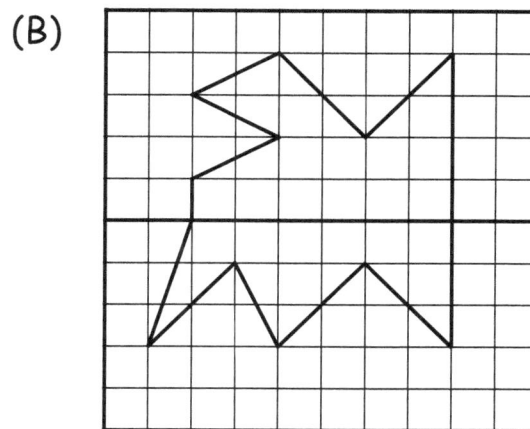

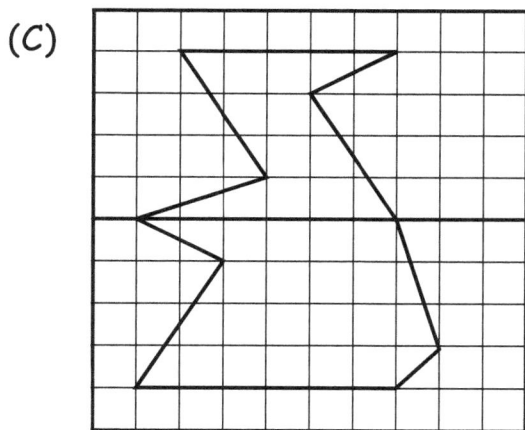

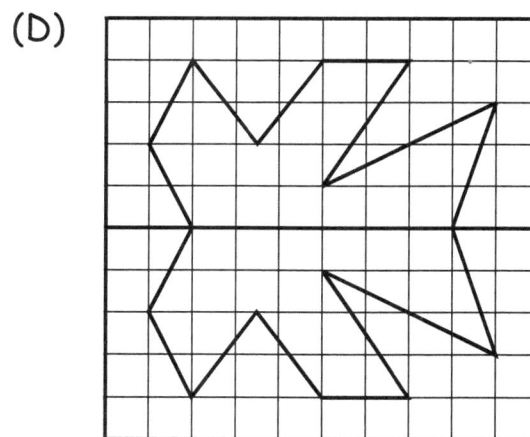

(A)

(B)

(C)

(D)

28. Mrs. John wants to change the carpet in her living room. What dimensions is she looking for ?

(A) Perimeter

(B) Area

(C) Volume

(D) Length

www.a4ace.com   www.math-knots.com

29. Milli's study room has an area of 45 sq centimeters and perimeter of 36 centimeters. What are the dimensions of the room ?

    (A) Length — 15 inches ; width — 3 inches

    (B) Length — 2 inch ; width — 13 inches

    (C) Length — 9 inches ; width — 5 inches

    (D) Length — 2 inches ; width — 16 inches

30. O is the center of the circle then XO is called ?

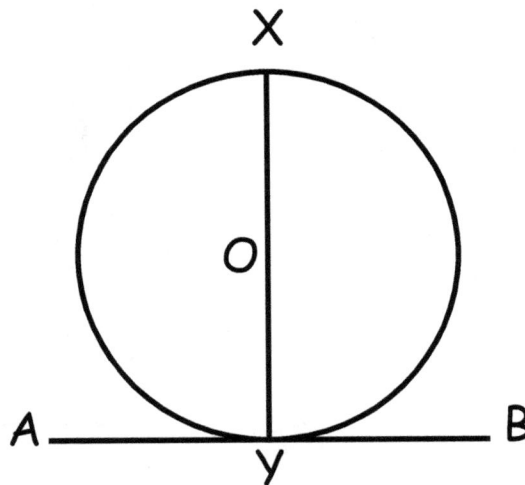

    (A) Diameter   (B) Arc

    (C) Chord   (D) Radius

31. which of the below triangle is a right angle triangle ?

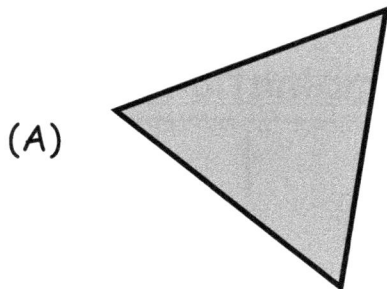

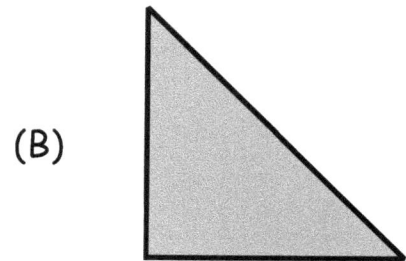

(A)

(B)

(C)

(D)

32. Calculate the mean of the below data set ?

| 25 , 7 , 16 , 13 , 11 , 25 , 15 |

(A) 15   (B) 14

(C) 16   (D) 12

33. Amount of syrup that an infant takes when he is sick will be in ?

(A) Ounces   (B) Quarts

(C) Liters   (D) Pounds

34. Mary has the below accessories to get ready for her birth day party. She chooses Blue hand bag. In how many ways can she choose her remaining accessories.

Accessories Combinations

Hand bag	Hair clip	Belt
Green Hand Bag Blue Hand Bag	Red Hair Clips Yellow Hair Clips	Red Belt Gold Belt

Based on the information in the chart, which tree diagram shows all of Mary's possible combinations ?

(A) Green Hand Bag ⟶ Red Hair Clips ⟶ Red Belt

(B)
Green Hand Bag → Red Hair Clips → Red Belt, Gold Belt; → Yellow Hair Clips → Red Belt, Gold Belt
Blue Hand Bag → Red Hair Clips → Red Belt, Gold Belt; → Yellow Hair Clips → Red Belt, Gold Belt

(C)
Green Hand Bag → Red Hair Clips ⟶ Red Belt; → Yellow Hair Clips ⟶ Gold Belt

(D)
Blue Hand Bag ⟶ Red Hair Clips → Red Belt, Gold Belt

35. Tina makes a spin wheel to choose her dress color for her Birthday party. What is the probability of choosing a White dress in one spin ?

(A) $\frac{1}{5}$ (B) $\frac{3}{4}$

(C) $\frac{1}{2}$ (D) $\frac{3}{5}$

36. Find the mode for the following data set of numbers ?

77 , 71 , 73 , 77 , 77 , 61 , 67 , 69 , 29 , 49 , 59

(A) 71   (B) 73

(C) 67   (C) 77

37. What is the variable in the expression 77 x - 19 ?

(A) x   (B) p

(C) 121   (D) -35

         www.a4ace.com   www.math-knots.com

38. Julian made the following list of all his math quiz scores.
98, 99, 96, 46, 46, 47, 47, 51, 52, 53, 57, 59, 59
Which stem-and-leaf plot correctly shows Julian's quiz scores ?

(A)

Stem	Leaf
7	0 , 7 , 6
8	5 , 6 , 8 , 8 , 9
9	11 , 8 , 8 , 9 , 6

(B)

Stem	Leaf
4	7
5	5 , 8 , 8 , 8 , 9
6	1 , 8 , 9 , 6

(C)

Stem	Leaf
4	6 , 7
5	1 , 2 , 3 , 9
6	1 , 8 , 9 , 6 , 7

(D)

Stem	Leaf
4	6 , 6 , 7 , 7
5	1,2,3,7,9,9
9	8 , 9 , 6

39. A stationary store sells "c" erasers in one box. Jia bought 6 boxes today.
How many erasers did she bought in total ?

(A) c + 6

(B) c - 6

(C) 6c

(D) $\dfrac{6}{c}$

40. Jack and Jill are playing a spinning game

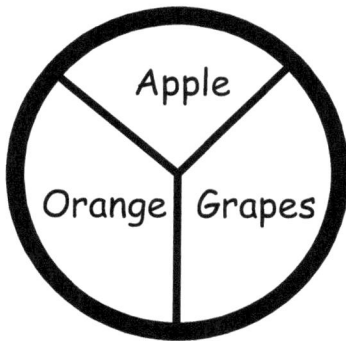

Spinner A

Spinner B

If Jack chooses Apples on the Spinner A. Show all possible outcomes in Spinner B in the form of a tree diagram.

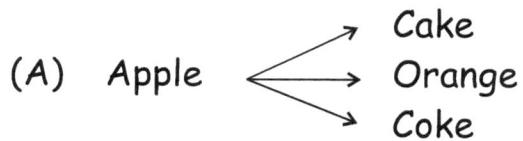

(A) Apple → Cake
→ Orange
→ Coke

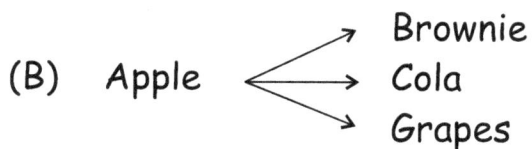

(B) Apple → Brownie
→ Cola
→ Grapes

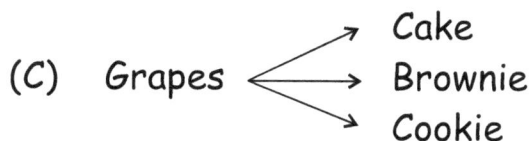

(C) Grapes → Cake
→ Brownie
→ Cookie

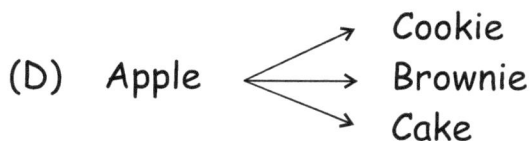

(D) Apple → Cookie
→ Brownie
→ Cake

41. The Grade 5 students solved various puzzles this week.
The below graph shows the number of puzzles solved by the students.

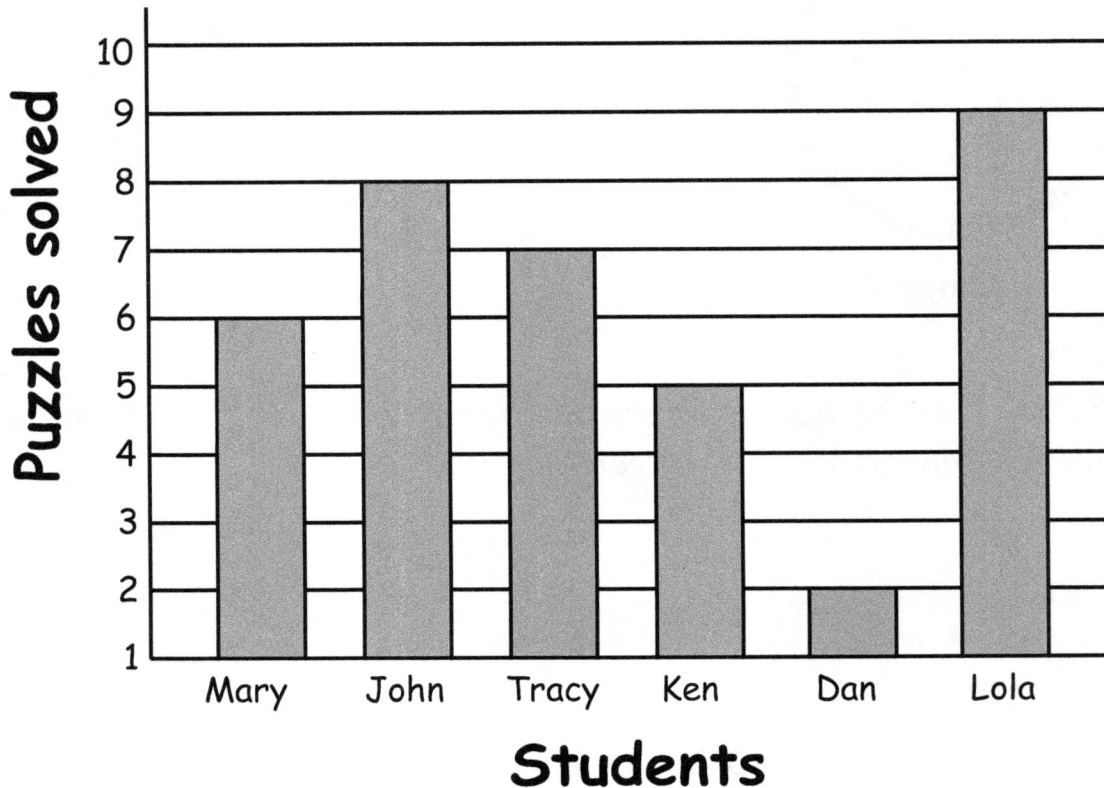

The greatest difference in the number of puzzles solved by which
of the below pair of students ?

(A) Mary and Tracy

(B) John and Lola

(C) Dan and Lola

(D) Ken and Tracy

42. Translate the below open sentence into words.

    9 - 5 = X

    (A) Lucy has 9 bags. she has sold 5 of them. How many bags are left with her now ?

    (B) Lucy had 11 bags. She bought 5 more. How many bags does she have now ?

    (C) Lucy had 9 bags. Her friend has 5 times more than her. How many bags does they have all together now ?

    (D) Lucy had 11 times more bags than her brother who has 5 bags. How many bags does they have all together now ?

43. Kate works on her math homework problems on Friday, only 7 more are left for her to work over the weekend. If total number of math problems are represented by m, which of the below expression displays the number of problems solved so far ?

    (A) $\dfrac{m}{7}$

    (B) m *7

    (C) m - 7

    (D) m + 17

44. Which of the below options represents a translation of seven units to right and three units to down ?

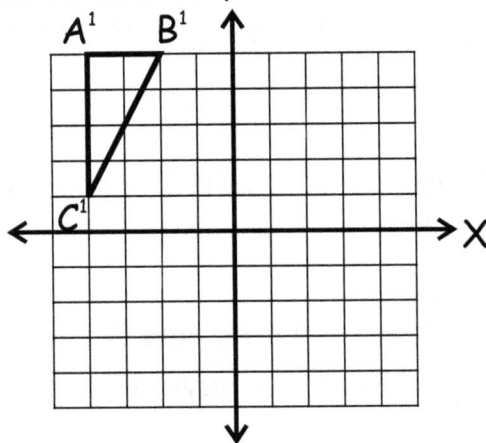

(A)

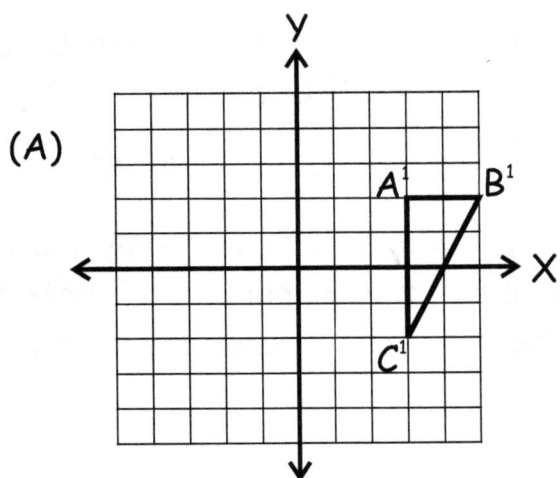

(B)

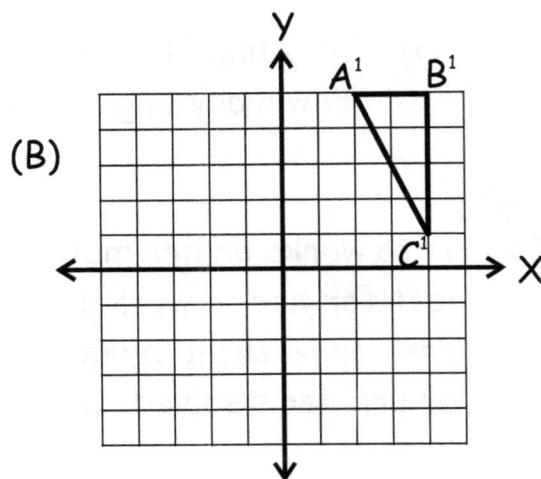

(C)

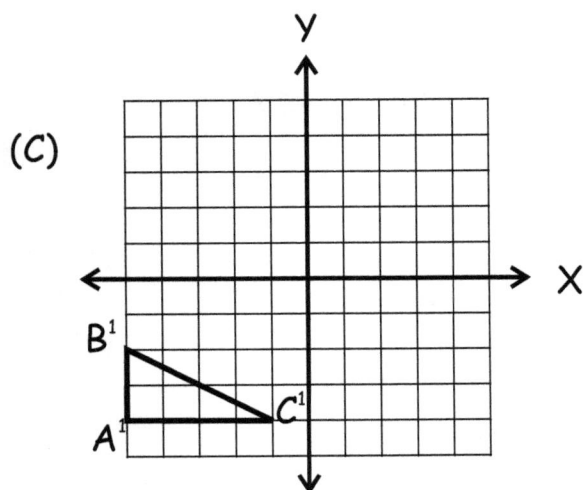

(D)

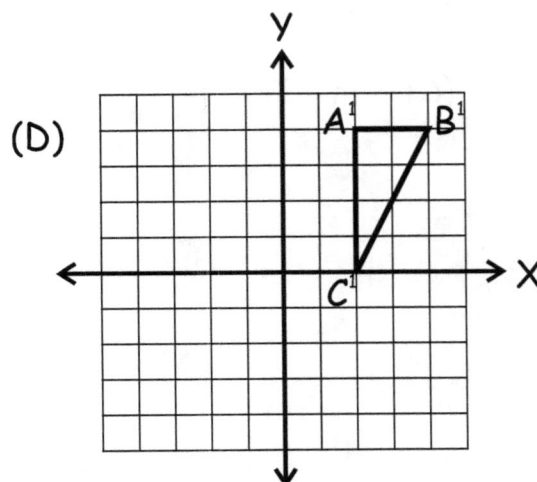

45. Henkel distributes 6449 blankets among 18 orphanages equally, representing the equation 6449 ÷ 18 = 358 R B. What does B represent in the equation ?

    (A) Number of blankets left
    (B) Number of Orphanages
    (C) Number of blankets each orphanage receives
    (D) Number of blankets each person receives

46. If "b" books are distributed among 24 kids in the class. Which of the below expression represents the number of books each student receives ?

    (A) b - 24   (B) b*24

    (C) b + 24   (D) $\frac{b}{24}$

47. Number sentence 3h - 2 = 12 can be written in words as

    (A) Bella baked 12 muffins in 3 batches. How many muffins did she bake in each batch ?

    (B) Bella baked h muffins in each of 3 batches and ate two of them. She now has 12 muffins left. How many muffins did she bake in each batch ?

    (C) Bella baked 3 muffins in 12 batches and ate 3 of them. How many muffins did she bake in total ?

    (D) Bella baked 3 muffins in 2 batches. How many muffins does she need to bake more ?

48. Danny solves five times as many problems as his friend Sia. Sia solves 12 problems. Which number sentence could be used to find m, the number of problems that Danny has ?

(A) $m = 5 - 12$

(B) $m = 5 \times 12$

(C) $m = 5 \div 12$

(D) $m = \dfrac{5}{12}$

49. This table below shows the soccer practice time by super team each day

Day	Time (hr)
1	0.45
2	1.00
3	1.15
4	1.30

If the pattern continues, What is the time spent by super team at the end of 5th day ?

(A)   1 hour, 55 minutes.

(B)   2 hours, 30 minutes.

(C)   1 hour, 45 minutes

(D)   1 hour

www.a4ace.com   www.math-knots.com

50. Find the fifth figure in the pattern ?

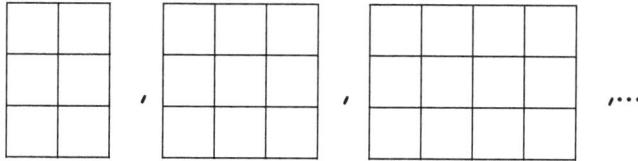

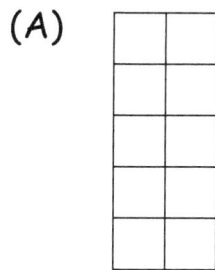

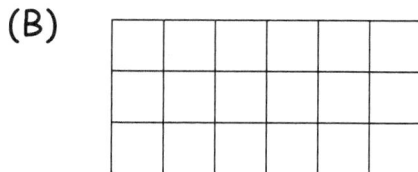

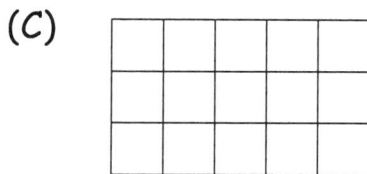

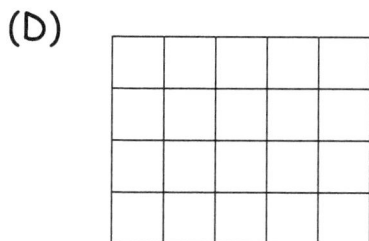

(A)

(B)

(C)

(D)

GRADE 5
SOL
Practice Test - 1
Answer Keys

Answer Key Test - 1

1. C

2. A

3. C

4. A

5. A

6. C

7. A

8. D

9. A

10. D

11. D

12. C

13. C

14. C

15. A

www.a4ace.com   www.math-knots.com

16. D

17. D

18. A

19. D

20. B

21. A

22. A

23. D

24. B

25. A

26. B

27. C

28. B

29. B

30. D

31. D

32. A

33. B

34. C

35. B

36. A

37. A

38. C

39. B

40. A

41. A

42. C

43. A

44. B

45. A

46. D

47. B

48. B

49. D

50. A

GRADE 5
SOL
Practice Test - 2
Answer Keys

www.a4ace.com
www.math-knots.com

www.a4ace.com   www.math-knots.com

Answer Key Test - 2

1. B

2. C

3. A

4. D

5. B

6. B

7. D

8. A

9. D

10. A

11. D

12. C

13. D

14. D

15. C

16. C

17. A

18. B

19. D

20. D

21. B

22. D

23. C

24. D

25. A

26. D

27. D

28. B

29. D

30. C

www.a4ace.com
www.math-knots.com

31. B

32. B

33. B

34. A

35. D

36. B

37. D

38. D

39. D

40. C

41. A

42. D

43. A

44. D

45. D

46. D

47. C (TRIANGULAR PRISM)

48. A

49. B

50. A

www.a4ace.com   www.math-knots.com

GRADE 5
SOL
Practice Test - 3
Answer Keys

Answer Key Test - 3

1. C

2. B

3. B

4. D

5. A

6. A

7. C

8. C

9. D

10. C

11. B

12. D

13. A

14. C

15. A

16. A

17. D

18. A

19. D

20. B

21. B

22. C

23. D

24. C

25. A

26. B

27. D

28. B

29. A

30. D

www.a4ace.com   www.math-knots.com

31. B

32. C

33. A

34. B

35. A

36. C

37. A

38. D

39. C

40. D

41. C

42. A

43. C

44. A

45. A

46. D

47. B

48. B

49. C

50. C

GRADE 5
SOL
Score Calculation

www.a4ace.com   www.math-knots.com

Score calculation

If you get this many times correct :	Then your converted scale scor is :
0	000
1	217
2	249
3	268
4	282
5	293
6	303
7	311
8	319
9	325
10	331
11	337
12	343
13	348
14	353
15	357
16	362
17	366
18	371
19	375

         www.a4ace.com   www.math-knots.com

20	379
21	383
22	387
23	391
24	395
25	399
26	403
27	407
28	411
29	415
30	419
31	423
32	427
33	432
34	436
35	441
36	445
37	450
38	456
39	461

40	467
41	473
42	480
43	487
44	496
45	505
46	516
47	531
48	550
49	582
50	600